# FIREWORKS!

## Pyrotechnics
## on Display

# FIREWORKS!

## Pyrotechnics on Display

## Norman D. Anderson

AND

## Walter R. Brown

ILLUSTRATED WITH

PHOTOGRAPHS AND PRINTS

DODD, MEAD & COMPANY · NEW YORK

## PICTURE CREDITS

Norman D. Anderson, 70; Eleutherian Mills Historical Library, 11, 14; Florida Division of Tourism, 2; *Harpers Weekly*, 6, 8, 12, 42, 43; Morgan-Nielsen Insurance Service, Remsen, Iowa, 37, 38, 39; Museum of the City of New York, 13; The New York Public Library Picture Collection, 9, 44; New York Pyrotechnic Products Company, Inc., 50, 51, 52, 53, 55, 56, 57; Parker Museum, Spencer, Iowa, 33, 34, 35; *Scientific American*, 17. All other photographs are by Todd Anderson. The print on page 10 is from *Pyrotechnia, On the Discourse of Artificial Fireworks* by John Babington, London: Thomas Harper, 1635. The print on page 19 is from *The Second Booke* by John Bate, London: Thomas Harper, 1635.

Library of Congress Cataloging in Publication Data
Anderson, Norman D.
Fireworks! : pyrotechnics on display.
Includes index.
Summary: Describes how fireworks came to be invented,
how they are made, the different kinds and their uses,
and some of the disasters they have caused.
1. Fireworks—Juvenile literature. [1. Fireworks]
I. Brown, Walter R., 1929–    . II. Title.
TP300.A5   1983        662'.1             82–45995
ISBN 0–396–08142–8

# Contents

1   Celebrating with Fireworks   7

2   Who Invented Fireworks?   15

3   Kinds of Fireworks and How They Work   20

4   Disaster in Iowa   32

5   Other Fireworks Disasters   41

6   How Fireworks Are Made and Displays Staged   49

7   Other Uses of Fireworks   63

8   Photographing and Enjoying Fireworks Displays   67

  Laws Governing the Sale and Use of Fireworks   75

  Index   77

*A fireworks display celebrated the inauguration of President Grover Cleveland in 1885.*

# 1
## Celebrating with Fireworks

WHAT CELEBRATION do you think of when someone says "fireworks"?

The Fourth of July, of course!

Fireworks have been used to help us celebrate our Independence Day since the very beginning of our country. Back in early July of 1776, John Adams wrote a letter to his wife Abigail. Adams, who later was to become the second President of the United States, had just signed the Declaration of Independence, perhaps the most important paper in our history. And, naturally, he wrote to tell his wife of the event.

"This day," he wrote, "will be celebrated . . . with parade, guns, bonfires, and fireworks, from one end of this continent to the other, from this time forward forever more."

And it has been.

But the people of the United States are not the only ones to use fireworks in their celebrations. And they were not the first. Special events in almost every part of the world have almost always included the setting off of fireworks.

*The opening of Victoria Bridge in Montreal, Canada, in 1860 was observed with an elaborate fireworks display.*

Fireworks—or pyrotechnics, as they are sometimes called—are used to celebrate the birthdays of many countries. The French Independence Day, which they call Bastille Day, always involves fireworks. In 1980, the King and Queen of Belgium visited the United States. A celebration in Washington, D.C., in honor of the 150th anniversary of the independence of Belgium, featured both fireworks and music.

But some of the other days that involve fireworks displays may surprise you. Did you know that the arrival of Spring is a time for shooting fireworks in India, Germany, and several other countries? Religious celebrations may also involve fireworks. Christmas Day is a

8

*Fireworks have been used at celebrations in countries all around the world. These were at Strassburg in Germany in 1744.*

time for shooting fireworks in many Christian countries. Some parts of the southern United States observe this custom. Holy Week in Mexico and the Day of Saints in South America always include fireworks in their celebrations.

Special events in Europe have long used fireworks as a part of the celebration. In 1613, fireworks were used to mark the marriage of Princess Elizabeth, the daughter of the King of England. The fireworks made up a huge display showing St. George on horseback, fighting an even larger, fire-breathing dragon.

The dragon was a part of most European fireworks displays for several centuries. It was usually made of a wooden framework cov-

9

*This huge dragon, with fireworks shooting out of its mouth, neck, and tail, was pulled along a wire.*

ered with "skin" of papier-maché. Rockets were attached to each rib of the terrible animal, and to its tail and mouth. The flaming monster could be floated in water, pulled along the ground on wheels, or fly through the air on wires. These figures were often shown in fierce, fiery battle with other dragons or with knights, such as St. George.

Many modern fireworks displays include *set pieces*, which are wooden frames with fireworks called *lances* attached. Lances consist of thin paper tubes filled with chemicals that give off a bright light when they burn. Lances that produce various colors of light are arranged to spell out words like "Fourth of July" or "Good Night." Other set pieces are designed to produce the outline of objects such as the American flag or a portrait of a famous person.

Displays of this type became common in England about three hundred years ago. And not only did these set pieces produce the

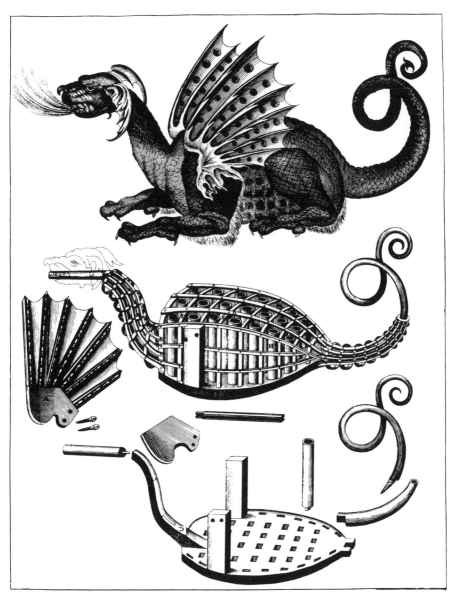

*This diagram from a book published in 1729 shows how a dragon set piece was constructed.*

*These set pieces showed firefighting equipment at a New York Fire Department celebration in 1885.*

outlines of various objects, but fireworks were added to represent flowing water or lava coming from the mouth of a volcano. One of the first permanent displays was set off every night as part of an English amusement park. It was called "The Forge of Vulcan." Vulcan was the Roman god of fire and the sparks from the burning fireworks were supposed to represent the fire in which the god worked with iron tools.

Many amusement parks in Europe still close nightly with this type of fireworks display. Sometimes these show exciting sea battles, the eruption of a volcano, or the historic burning of a city.

In the United States, fireworks have been used for many different events. The Philadelphia Centennial Exhibition in 1876 used many fireworks to celebrate our country's 100th birthday. The Brooklyn Bridge in New York City was opened in 1883 with an hour-long

*The opening of the Brooklyn Bridge in New York City was the occasion for an elaborate fireworks display in 1883.*

display. Fourteen tons of rockets and flares were set off from the center span and the tops of the bridge towers. And the Hudson-Fulton celebration in New York in 1909 included a spectacular display of fireworks. This event was in honor of the 300th anniversary of Henry Hudson's discovery of the Hudson River and the 102nd anniversary of Robert Fulton's first successful steamboat trip on the river.

The opening of the Erie Canal was announced in New York City by 1,500 fireballs and 374 rockets crisscrossing in the sky. Sheets of "gold and silver rain" fell from them and into the waters of the Hudson River. This was in 1825.

The Panama-Pacific Exposition in San Francisco in 1915 had an unusual fireworks display. Here 48 searchlights threw colored beams of light on smoke made by bursting fireworks and plumes of steam.

13

*Celebration of the opening of the Erie Canal was climaxed in New York City with a "magnificent and extraordinary" fireworks display atop the City Hall on November 4, 1825.*

The World's Fair of 1982, held in Knoxville, Tennessee, was ended each night by another new display. It was a combination of fireworks exploding in the sky and a display of laser beams. The blue and red laser beams played across the smoke left by the exploding fireworks, creating a spectacular show of color.

# 2

## Who Invented Fireworks?

No ONE REALLY KNOWS who was the first person to make fireworks.

It probably happened by accident many, many hundreds of years ago. We don't even know for certain in which part of the world fireworks were first made. But most people think that the discovery was made in China.

Fireworks are made of gunpowder. Gunpowder can be made by mixing ground-up charcoal with chemicals called sulfur and saltpeter. Both of these chemicals are found almost everywhere.

Charcoal is produced when wood burns without enough oxygen for it to be turned into ashes. This can happen near the bottom of any open fire. The incompletely burned wood—the charcoal—burns very slowly and produces a great deal of heat. Perhaps you have used charcoal to cook food at a picnic.

We can guess how the discovery of how to make gunpowder came about. There may have been a group of people living in China who did not have regular salt to put on their food. Without regular salt, they may have seasoned their food with saltpeter which has a salty taste. There are still people in China today who use saltpeter in their cooking.

While adding saltpeter to cooking food, someone may have spilled a little into the fire. The people would have noticed that the saltpeter made the fire burn more brightly.

Once this discovery was made, these ancient people might have decided to mix charcoal with saltpeter intentionally. This would make a mixture that would continue to burn while it was being carried from place to place. This would allow them to start a new fire easily when they moved to a new camp.

It is not hard to imagine that some yellow sulfur became accidentally mixed with the burning charcoal and saltpeter. Sulfur is found commonly in rocks all over the world. Perhaps a rock containing sulfur was used to line a fire pit. Gunpowder would have been discovered—with a bang!

Or it may be that gunpowder was first discovered in China by an ancient scientist called an *alchemist*. The science of chemistry began with experiments by alchemists. These people believed that they could turn common materials into valuable gold and silver. Many of the early experimenters worked a great deal with sulfur because it has a color similar to that of gold. They may have tried to turn saltpeter into silver. And they undoubtedly used charcoal as a fuel for the fires they needed to heat their mixtures. It may be that an alchemist tried mixing all three and accidentally made the first gunpowder.

One of the first written records of the use of fireworks is Chinese. A description of how to make gunpowder into a small, firecracker-like "fire pill" was found in a book that is believed to have been written in A.D. 1040. According to other Chinese writings, fireworks similar to those we use today were probably used in China before the year A.D. 1175.

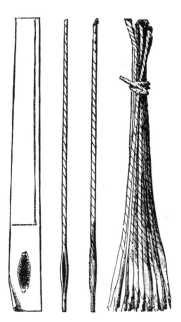

*These early Japanese parlor fireworks were made by placing chemicals on one end of a paper strip and twisting it.*

Many people believe that the famous traveler Marco Polo brought gunpowder back to Europe when he returned from his visits to China and India in the late 1200s. Other people think that gunpowder may have been invented in Europe by people who did not know about the discoveries of the Chinese.

The Greeks and Romans used fire as a weapon as early as the year A.D. 670. This weapon was called "Greek fire" and contained sulfur and saltpeter, as well as several other chemicals. The mixture was a liquid and could be poured or thrown at the enemy as it burned. Therefore, it was also known as "liquid fire."

Sometime later—no one is certain exactly when—gunpowder was used to throw "Greek fire" greater distances. In this way, the first guns were invented. However, instead of firing bullets, they "fired" the burning liquid. So these weapons may have been the first true fireworks.

The first use of gunpowder in a real gun was probably in Europe. The first cannon is thought to have been developed in the 1300s in what is now Germany.

As you can see, the development of fireworks goes along with the development of weapons for war. The first display of fireworks for entertainment in Europe was put on by military men who knew how

*Novelty fireworks of today. Sparks shoot out from the cannons of this ship and tank. Propeller fireworks spin rapidly and the blades pull them upwards, leaving a trail of sparks and smoke behind.*

*A "Green Man" dressed so that he could not be easily seen as he moved about in the darkness setting off fireworks with his torch.*

to handle explosives. They were hired to set off explosions of bombs and to fire rockets and mortars to celebrate important events.

By the early 1500s, some people who were not in the army were making a living by setting off fireworks to entertain audiences during the celebration of special events. These people employed carpenters, painters, plasterers, and masons to build displays. They made their fireworks themselves and then hired men to set them off. These men wore green uniforms and put green leaves in their hats so that they could move around inside the display without being seen as they lit the fireworks. Because of their costumes, they were called "Green Men." One of the first large displays of fireworks in England put on by one of these commercial groups was in 1533. Its purpose was to greet a new Queen, Anne Boleyn, when she arrived in London.

# 3

# Kinds of Fireworks
# and How They Work

FIREWORKS HAVE CHANGED a great deal since their invention 1,000 or so years ago. They have become more powerful and in many ways more dangerous. Also, the people who make fireworks are always designing new kinds and working on ways to make the old favorites more spectacular.

One of the most popular home fireworks is the *sparkler*. This is a piece of metal wire coated with a mixture of chemicals. When the sparkler is lit with a match, it gives off a shower of white or colored sparks. Sparklers also give off considerable smoke and heat. The chemicals in sparklers burn at a temperature of about 1650° F. Because of this, they are dangerous to use without close supervision. As you will read later, it was a sparkler that caused one of the worst fireworks disasters in the United States.

Another popular home fireworks is called *snakes* or *worms*. A small packet of chemicals wrapped in cardboard or foil is lighted with a match. As the chemicals burn, a long, gray ash is produced.

*Sparklers come in different colors.*

Because this "snake" is larger than its container, it is slowly pushed out. It looks very much like a snake crawling out of a hole.

Almost all other fireworks are classified as "explosive fireworks." They are all made pretty much alike. They contain a chemical called an *oxidizer*. The purpose of this chemical is to produce the oxygen needed for rapid burning.

Of course, all explosive fireworks contain a *fuel*. This chemical combines with the oxygen in the oxidizer. This process is called

21

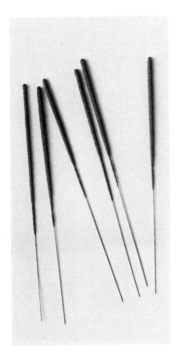

*When lit, sparklers give off a shower of burning sparks.*

burning, and large amounts of heat and light are given off very rapidly.

The fuel in fireworks is a solid and must be ground until it is exactly the right size and shape. As you might guess, this grinding process is a very dangerous part of fireworks manufacturing. However, if it is not done correctly, the fireworks will not explode in the way they were meant to.

Another important part of explosive fireworks is called the *binder*. The fuel must be carefully mixed with this material in such a way that the tiny fuel particles are exactly the right distance apart. If this mixing is not done correctly, the fireworks may explode too soon, or not at all.

*Snakes are produced by burning chemicals that create a long, gray ash.*

Some fireworks contain chemicals that add special effects to the explosion. Because of these special effects materials, fireworks may produce colors, sparks, smoke, noise, or a combination of these.

All of these chemicals are tightly packed into a paper or cardboard container, along with one or two fuses. The final form of many fireworks is spherical or in the shape of a cylinder. The length of the fuses controls when different parts of the fireworks explode, and the shape of the container controls the shape of the bursts. The

*These 1½-inch firecrackers are the largest size that can be legally sold in the United States.*

huge, round displays of Japanese fireworks are produced from ball-shaped fireworks.

Because fireworks are explosives and are dangerous, their manufacture and sale in the United States are controlled by the Federal government. There are several chemicals that cannot be legally used in fireworks at all, and others that cannot be used together in the same fireworks. The purpose of these government regulations is to try to prevent injuries caused by the manufacture and use of fireworks.

Class C fireworks are the ones sold in some states for home use. The Federal government requires that these explosives be carefully labeled and not be too big. Class C explosives cannot contain more

*Smoke bombs give off a cloud of smoke, but do not explode. Other bombs are powerful and dangerous explosives.*

than 50 milligrams of explosive powder. This is a very, very small amount—only about 1/500 of an ounce. Even so, these types of fireworks cause many, many injuries each year. Several states will not allow even these Class C explosives to be sold or set off by the public.

Unfortunately, there are people who make and sell fireworks illegally. Since they are already breaking the law, these people often put more explosives into their fireworks than the law allows. *Cherry bombs*, the *M-80*, and the *Silver Salute* are examples of these powerful, dangerous, and illegal fireworks that are sold in some places.

Some illegal fireworks have been found to contain thirty to forty times as much explosive powder as the law allows! During the past few years, more than one-third of the Fourth of July fireworks injuries were caused by these illegal fireworks. And nearly all of these injuries were so serious that the victims had to be taken to hospitals for treatment.

Because of government regulations about the sale and shooting of fireworks by the public, most of our present-day celebrations use displays put on for the enjoyment of thousands of people at one time. They are set off by experts—modern Green Men—who are well trained in how to handle explosives safely. Because of this, we can enjoy fireworks that are larger, brighter, louder, and safer than they were only a few years ago.

Now let's look at some of the different kinds of explosive fireworks. *Roman candles* are long tubes that shoot a series of flaming stars into the air. The stars are hard, tightly packed balls of chemicals that burn as they fly. They are often made to burn with bright colors and to throw sparks in all directions. Inside the Roman candle tube, the stars are packed one on top of another. Between each star is a layer of powder that, when it explodes, throws the uppermost star out of the tube.

*Rockets* are different from Roman candles, even though they look something alike. The rocket tube flies through the air in the same way a spaceship flies through space. The gunpowder inside the rocket burns rapidly. This produces a gas that pushes against the walls of the tube in all directions and escapes from the rear. This gas pushes the rocket tube high up into the air. The rockets may carry stars, or they may contain bombs, parachutes, or paper floats. A second, slow-burning fuse can be used to make the rocket explode and throw out these special effects near the peak of its flight.

*Bombs*, ordinary *firecrackers*, and *flash crackers* are all made in much the same way. The gunpowder is tightly packed inside a card-

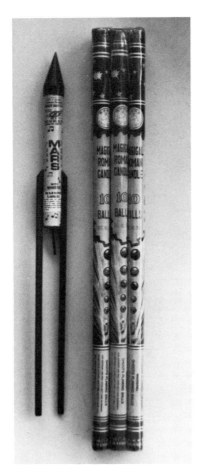

*Popular fireworks include the rocket (left) and Roman candles.*

board container. When it is lit, the powder burns very rapidly. This burning produces a gas that pushes out in all directions, just as in the rocket. However, unlike the rocket which is open on one end, the crackers or bombs are tightly sealed all around the powder. The gas made by the burning gunpowder blows the container apart with a loud noise.

*Fountains give off a spray of colored sparks that shoot up from the ground. They reach several feet into the air.*

*Fountains* and *gerbes* are fireworks that throw colored sparks into the air from the ground. They are similar to rockets except that the open end is pointed upward. Instead of driving the container upward, the burning gunpowder is forced up and out of the tube or container.

*Wheels* or *pinwheels* are rockets that are attached to a solid, upright frame. Instead of flying off into the air, these fireworks spin around and around the nail that is holding them to the frame. Several rockets may be attached to a larger wheel and cause the entire mechanism to go around.

*Shells* are explosive fireworks that are shot into the air from mortars. A mortar is a simple firing device made of a heavy cardboard or metal tube that has been sunk into the ground or is held securely by some other means. The shell comes wrapped in heavy

*These shells with fuses are ready to be placed in the rack of mortars behind them.*

*A burst of bright light is produced by an exploding shell.*

paper and is lowered into the mortar with the fuse extending from
the top. When the fuse is lit, the gunpowder charge in the bottom
of the shell explodes and the shell is propelled upward. A slow-
burning fuse inside the shell causes it to explode near the high point
of its flight into the nighttime sky. Shells may be as large as *two feet*
in diameter, and carry hundreds of stars and other displays into
the air.

Some shells are called "announcers" and are used to tell people

30

that the fireworks show is about to begin. The noise of the bursting shell can be heard up to ten miles away!

Other types of shells may produce shining clouds filled with masses of silvery sparks. Or the shell may explode and release stars and sparks glowing with almost any color. One spectacular shell explodes to release first a green mist and then a silver cloud. Other shells throw dozens of colored comets that dash brightly toward the earth. These different kinds of shells are the main attraction at most fireworks displays.

Wheels, gerbes, fountains, and other types of fireworks can be combined to make special picture displays. By mounting the fireworks on a framework, and setting them off in the right order, a colored picture of the flag, the Statue of Liberty, or the Liberty Bell can be made. One display even shows two glowing spaceships firing rockets at each other!

# 4

# Disaster in Iowa

THE 5,000 PEOPLE who lived in Spencer, Iowa, in late June of 1931 were busy preparing for their annual Fourth of July celebration. A parade was being planned, picnics were being prepared, and several of the stores on the main street of the small town were full of fireworks.

One of these stores was Bjornstad's drugstore. The display of fireworks, rockets, and pinwheels was in the basement. Only one clerk was assigned to help customers there, since most of the people who lived in Spencer knew all about the fireworks and could find what they wanted without assistance. On this Saturday morning, two small boys were the only customers. Eleven-year-old Billy Kilpatrick reached into an open box of sparklers.

"Is this a punk?" he asked his friend. Punks were slow-burning fuses on wires, and looked very much like sparklers. They were lit with a match and then could be carried around for several minutes and used to light other fireworks.

The other boy wasn't sure whether the object was a punk or not.

*The Rexall drugstore and several other buildings across the street from where the fire started in Spencer, Iowa, were soon engulfed in flames.*

The clerk in the fireworks department was busy setting out more fireworks and apparently did not hear the question.

"Let's light one and find out," one of the boys suggested.

A match flared and was touched to the end of the sparkler. A bright spray of sparks suddenly erupted. The alarmed clerk shouted, and Billy dropped the glowing sparkler—right into a box of large cherry bombs.

As the cherry bombs began to explode, the boys and the clerk rushed for the stairway. They knew there was nothing that could be done to stop the fire. As they reached the main floor of the drugstore and shouted a warning to the customers and clerks there, the whole basement seemed to burst into flame. Dense smoke filled the

*The telephone operators on the second floor of this building stayed at their switchboards as long as they could.*

shop and those people farthest away from the door had to drop to the floor in order to breathe.

The fire quickly spread. The Clay County Bank was soon ablaze, and tellers gathered up all the cash and records they could and ran for their lives, leaving a million dollars in cash locked in the vault. The opera house, the American Legion hall, a clothing store, and a second drugstore quickly caught fire.

The Spencer Volunteer Fire Department, working with only a few pieces of equipment, soon realized that they could do little to control the spread of the fire. The operators at the telephone company stayed at their switchboards as long as they could, sending calls for help to neighboring towns. The last message to leave Spencer was a telegram directed to the newspaper in Des Moines, Iowa. The message read: "Town is burning. Send plane with dynamite and fire chief. Water pressure is gone."

This telegram shows how hopeless the people of Spencer had become. Most of the buildings in the heart of the city were on fire. A brisk wind pushed the flames from building to building. Stocks of fireworks on display in many of the stores exploded, showering the nearby buildings with sparks. The flames were so hot that the pavement of Main Street buckled and cracked. Records and the stock of many businesses were being destroyed along with the valuable buildings.

Three airplanes from Des Moines arrived in response to the tragic telegram that had been sent hours earlier. On board were the fire chief of the city of Des Moines, several of his experienced firefighters, and two explosives experts. They quickly went to work, planting dynamite under buildings that had not been reached by the flames. The destruction of these buildings cleared a wide path in front of

*Ruins of the Clay County Bank (on the right) and the Bjornstad drugstore (leveled area in the center of the photograph).*

35

the advancing fire. Without fuel to burn, the fire finally died down. By the time the sun had set, the fire was under control.

As the fire cooled, Spencer began to count its losses. It was very fortunate that no one had been killed in the fire. But nearly seventy-five businesses and offices were destroyed or seriously damaged. And many valuable records were lost forever.

But there were some happy moments. One of the most worried persons in town was the president of the bank. It took ten days for the ruins of the bank to cool enough for anyone to try to get to the vault. When the vault was finally opened, the million dollars was found to be undamaged.

The Bjornstads rebuilt their drugstore, and Main Street slowly became lined with new, modern buildings. Billy Kilpatrick, who was never blamed for his part in the fire, grew up and became a captain in the Air Force during World War II. The scars of the disaster gradually disappeared.

But the memory of the terrible day remained. Some of the town's leaders decided that no other community should have to suffer as

*Bjornstad's drugstore as it appears today.*

their town did. They got together and passed a resolution suggesting that the government of the United States should prohibit the sale of fireworks to the public. People could still enjoy fireworks, they argued, if they were set off by professional, licensed operators without the dangers of home use.

This was a fairly new idea, and it did not impress many people. The state of Michigan did have a law such as this one, but no other state had agreed that such a law was worthwhile. Even the lawmakers in Iowa were slow to pass such a law in their state. They discussed the idea for five years, and still might not have acted if disaster had not struck two other Iowa towns.

In 1936, the town of Remsen had a population of only about 1,200 people. On the Fourth of July of that year, a small girl was playing with a sparkler near a tent that had been set up for the annual celebration. The flying sparks frightened the child, and she dropped the sparkler. The burning fireworks fell into a pile of oil-soaked rags and the fire quickly spread to the tent. From there, the fire leaped to the walls of a nearby garage, and was soon raging out of control.

*People watched helplessly on Remsen's flag-lined main street.*

*The fire in Remsen was soon out of control and firemen had to pull back. Moments later some of the wooden buildings collapsed.*

*A fireman sprays water between two buildings in an attempt to keep the fire in Remsen from spreading.*

Only a few minutes later, another fire started in the little community of Oyens, which stood about ten miles west of Remsen. This fire started in a vacant lot near the Oyens Lumber Yard. No one will ever know exactly how this fire started, but many people suspected that a firecracker had been thrown into the dry grass that covered the vacant lot.

Oyens was a very small town, with probably not more than a hundred people living there. Therefore, it had no fire department. A call was made to Remsen, but the fire department there was busy fighting the fire that was rapidly destroying the town. Calls were made to several other nearby towns, but all the equipment and men

39

had been sent to Remsen. The people of Oyens could do little except watch their homes and businesses burn.

In Remsen, it took six hours and dynamite to stop the flames. By the time this was done, half the town was gone. More than a hundred people were homeless and twenty buildings were burned out.

At last, the General Assembly of Iowa acted on a bill to forbid the sale of fireworks in the state. The law, which went into effect in 1938, also made it illegal for anyone to set off any fireworks without a permit. Since that time, several other states have used the Iowa law as a pattern for laws of their own.

# 5

# Other Fireworks Disasters

THE FIRES IN IOWA were important because they resulted in laws that controlled the sale and use of fireworks. But they were neither the first nor the last of the terrible disasters to be caused by fireworks.

Perhaps one of the most terrible fires caused by careless use of fireworks occurred on July 4, 1866, in Portland, Maine. The city was decorated for the celebration of Independence Day with flags and streamers. Crowds of people watched a long parade that wound its way through the streets. Children played as they waited for the parade by tossing firecrackers here and there.

One of these lighted firecrackers was thrown over a fence and into a boatbuilder's yard. It exploded in a pile of wood shavings and a fire started. The fire was small, and no one noticed it until it started to spread to nearby buildings.

All of the buildings in the area were made entirely of wood and a stiff wind from the south pushed the flames from roof to roof. The fire jumped quickly across streets, moving from block to block, and by dark had reached the heart of the city. By the time the fire burned

*The ruins of the post office and surrounding buildings in Portland, Maine, in 1866.*

itself out, the city was a charred ghost town. Not only was the business district gone, but more than 10,000 people were homeless.

Of course, there are disasters caused by fireworks that are not part of a Fourth of July celebration. As noted earlier, fireworks are used all over the world in celebrations of all kinds. Such celebrations do not always go as they are planned.

In 1749, the major cities of Europe were celebrating the peace treaty that had been signed the year before at Aix-la-Chapelle. This treaty was important to most of the countries of Europe because it marked the end of many separate wars. Britain had been at war with Spain since 1739. When France entered this war on the side of Spain, Britain attacked the French armies stationed in the New World. Later, in 1741, the countries of Prussia, Saxony, Bavaria, and France

*Many of those left homeless by the Portland fire had to live in tents until new houses could be built.*

attacked Austria. England immediately entered this war on the side of Austria.

These wars continued until 1748. No one really won any of them, but all of the countries involved lost many men and ships, and used up much of their supplies of weapons. Tired and discouraged, they finally agreed to sign a peace treaty. The wars were to start up again within a few years, but in 1748 everyone hoped that the treaty was the end to the terrible conflicts. Every country that had been involved in these wars decided to celebrate the end of the death and destruction. Most of these celebrations involved the use of fireworks.

For the celebration in London, the Ruggieri brothers, famous for their fireworks, were brought from Italy. They worked for five months to build their display, which was made of lumber and white canvas.

*The fireworks display in London in 1749 in honor of the peace treaty of Aix-la-Chapelle resulted in the fire shown at right.*

When they finished, they mounted 10,650 rockets, pinwheels, and firecrackers on the framework. The whole display was 410 feet long and more than 100 feet high!

The composer George Frederick Handel wrote a special overture for the event which he called *Music for the Royal Fireworks*. In addition to the fireworks and music, 100 cannon were to be fired.

Everything was finally ready for the exciting evening. The fireworks were mounted. The orchestra was tuned and in its place. The cannon were loaded and ready to fire. But the Ruggieri brothers were having an argument. Some of them wanted to use an old-fashioned match-type fuse to light the fireworks. Other members of the family wanted to set off the display with a trail of gunpowder poured on the ground. It was finally decided to try the gunpowder trail. But sparks from the burning powder set off all the fireworks at the same time and started a fire that burned part of a nearby theater.

A worse disaster happened in the city of Paris during the celebration of the same peace treaty. Two separate fireworks displays had been planned—one by a French fireworks company and another by a group of Italian experts. Again, an argument broke out. Each group thought it should have the honor of showing off its display first. Since they could not agree on who should go first, they decided to set off both displays at the same time. Somehow, an explosion took place and forty people were killed.

The United States has had its share of disasters during large fireworks displays. One of the worst of these happened in November, 1902, in New York City. A man named William R. Hearst had been elected to Congress and a huge fireworks display was planned to celebrate his election. As part of the show, sixty huge shells, each several inches across, were to be fired into the air.

These types of shells were, and still are, fired from hollow tubes

*Picture postcards that were exchanged on the Fourth of July in the early 1900s usually featured a firecracker.*

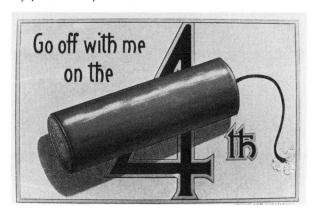

*These racks of mortars are ready to receive their shells.*

*Fuses connecting the shells in a rack extend from the tops of the mortars and allow several shells to be fired in rapid succession.*

called mortars, which are usually sunk into the ground. In the 1902 display, the mortars were simply set on the ground, pointing upward. When the first mortar was lit, it somehow exploded without sending its shell into the air. When the shell exploded inside the mortar, it blew the metal tube to bits. The blast knocked over the other mortars and fire from the first explosion set them off. The huge shells were fired directly into the crowd of people waiting to see the show. Fifteen were killed and eighty others were seriously injured.

In recent times, the United States government has issued several regulations dealing with the manufacture of fireworks and limiting the size of fireworks that are sold to the public. Most of the states also have laws governing the sale and use of fireworks. An appendix

*In states where fireworks can be legally sold, fireworks stores are a common sight along the highways.*

at the end of this book contains a list of the states and the status of fireworks sales and use in each.

An annual study made by the Consumer Product Safety Commission indicates that there are approximately 11,000 fireworks-related injuries in the United States each year. A majority of these injuries are caused by the misuse of fireworks, rather than by fireworks that do not work properly. Most of these injuries are caused by firecrackers, and the use of oversized illegal fireworks continues to be a problem.

# 6

# How Fireworks Are Made
## and Displays Staged

THE MANUFACTURING OF FIREWORKS in most parts of the world is a family business. The way the explosives are ground, mixed, and packed are secrets that have been passed down from generation to generation, sometimes for centuries. One major fireworks manufacturer, the Ogatsu family of Japan, has been in business for nearly four hundred years. In this country, the Grucci family of Bellport, New York, a small village on the southern shore of Long Island, has been in the fireworks business for five generations. They are often called "the first family of fireworks" in the United States.

Most of the fireworks sold in the United States are imported from overseas, primarily from China, Taiwan, Japan, and South Korea. However, there are a couple of very large and several smaller fireworks manufacturing companies in the United States. Some of these firms make only the smaller Class C fireworks. A few, including Grucci's New York Pyrotechnic Products Company, make the shells and other fireworks used in the large displays.

*Felix Grucci gives directions to sons James (left) and Felix, Jr. (right) as they load mortars for a fireworks display in New York City.*

The Grucci family has had the honor of furnishing the fireworks for perhaps our best-known display, which takes place each Fourth of July at the Washington Monument in our nation's capitol. They also staged the twelve nights of illumination for the 1980 Winter Olympics in Lake Placid, New York, and often supply the fireworks displays at New York Mets baseball games and the Boston Pops Orchestra concerts.

In 1979, the Gruccis represented the United States at the annual International Fireworks Competition in Monte Carlo. The preparations took six months and seventy-four-year-old Felix Grucci, head of the family, personally supervised every phase of the fireworks being made and set off. During the five-minute finale at Monte Carlo, 1,500 shells were fired electrically in rapid order.

Grucci recalls, "People ran out of the Casino because they thought the place was being bombed." A passing ship radioed from far out in the Mediterranean Sea, "Is Monte Carlo on fire?" And when the show ended with a thundering burst of 8-inch shells, the honking of car horns and the bellowing of foghorns from yachts in the harbor told the Gruccis that the display had been a big success. For the first time, an American firm had won the competition.

In the past, there was little improvement in fireworks from year to year. But recently, scientists have become more interested in the making of fireworks. Their main goal is to make fireworks that are safe to use, but because of their research, new ways of making more spectacular explosives also are being discovered. Many new colors

*Felix (left) and other members of the Grucci family pose with an electrical panel used to set off a large display such as the one they staged in Monte Carlo.*

*Felix, Jr. (left) and James Grucci display two 8-inch shells that rise to a height of 600 feet and produce dazzling patterns of color.*

that last longer before they burn out have been developed in the last few years. And new methods of setting off fireworks, using electricity, have made displays not only safer but also much more spectacular.

The making of fireworks has always been a dangerous job. The explosive powder has to be ground, mixed with binder, and the fireworks have to be packed tightly in paper containers. An explosion can take place at any time.

One of the worst disasters in a fireworks factory in the United States was in April of 1930. It was at the factory of the Pennsylvania Fireworks Display Company, about sixteen miles from Philadelphia.

The factory was made up of eight small buildings scattered over

52

*The various ingredients found in a medium-sized shell.*

about seven acres of land. At about ten o'clock in the morning, the president of the company heard a small explosion that he thought sounded like a firecracker. Almost immediately afterward, the shed in which the gunpowder was stored blew up with a terrible roar. Apparently the blast from the explosion set off the fireworks being assembled in the factory's main building near the center of the plant. Ten persons were working in this building and all of them were killed.

This blast was quickly followed by another explosion and a fire that destroyed all of the buildings of the fireworks plant. Windows for miles in all directions were broken by the shock waves. Automobiles passing the plant on a nearby highway were overturned or shoved off the road. Blocks of cement and tile from the factory buildings were thrown for hundreds of yards. Dozens of people suffered cuts and bruises from the flying debris.

Modern fireworks factories do several things to avoid catastrophes such as the one in Pennsylvania. All chemicals are kept separate from each other, usually in separate buildings, until the last stage of putting the fireworks together. Each of the buildings is usually constructed in such a way that the walls will easily fall away from the blast without allowing the roof to fall. There are many escape exits around the buildings, so that if a fire does start, the workers have a chance to escape.

If you were to visit the Grucci plant, you would find a three-room office and a row of widely spaced sheds and outbuildings. In the office various members of the Grucci family work on plans for future fireworks displays. Others may be designing new kinds of fireworks or looking for ways to make them safer. It was the Gruccis who developed a shell that did not need a string wrapping to hold it together. The string on the older type of shell sometimes caught on fire, taking a piece of the burning shell down with it. This burning debris was a fire hazard and there also was the danger of it injuring spectators.

*Some of the small buildings making up the Grucci fireworks factory in Bellport, New York. During nice weather newly made fireworks are dried on the tables in the foreground.*

Before entering a small black powder house some distance from the office, you have to rub your hands on a grounded copper plate mounted on an outside wall. This is done to remove any static electricity from your body and clothing. No clothing made of synthetic materials, which easily produces sparks, is allowed inside. And of course no smoking is allowed anywhere near the explosive fireworks.

Once inside, you see a sign that reads, "Do not use ferrous tools to open containers." The reason is that a spark from an iron tool could easily set off an explosion. Another sign says, "Not more than 2 regular employees and not more than 25 pounds of Class 9 and 10 pyrotechnics allowed." This rule is to prevent more than two people from being injured if there is an accident and to keep any explosion as small as possible. The Gruccis have never had an employee fatally

*A workman at the Grucci fireworks factory making salutes, which are the noisemakers at a fireworks display.*

injured in an accident at their plant and they do everything possible to keep it from ever happening.

There are hundreds of fireworks displays around the country each Fourth of July and on other special occasions. The fireworks companies have only enough regular employees to stage the larger ones. The others are put on by part-timers who enjoy shooting fireworks and who have learned how to do so safely. To learn how one of these smaller displays is staged, let's go behind the scenes at the

*An employee in the final steps of making salutes.*

*Bill Shearon lowers a steel mortar into a hole. The next step will be to pack dirt around it to hold it firmly in an upright position.*

1982 Fourth of July celebration at the North Carolina State Fairgrounds in Raleigh.

The display is sponsored each year by the Raleigh City Parks and Recreation Department at a cost of about $6,000. The fireworks for the 1982 celebration were purchased from the Vitale Fireworks Manufacturing Company of New Castle, Pennsylvania. Some were made in this country and some were imported from China and Japan. They are shot off from the field inside the racetrack in front of the grandstand—a place where thousands can watch in safety.

Bill Shearon, Jr., of the nearby town of Wake Forest, works part-

*Racks of cardboard mortars are held upright by wooden stakes driven into the ground.*

time for Vitale and had the job of setting up the display and lighting the fireworks. Bill learned how to handle explosives in the service and has attended special schools on how to handle fireworks. Some states require that those setting off fireworks have a certificate, which is obtained by attending a school and passing a test.

Bill, his fourteen-year-old son Chris, and two other helpers arrived at the fairgrounds shortly after lunch. In the back of their truck were shells, mortars, set pieces, and the tools they would need. One of the first jobs, and a hot one, was to dig holes for the metal tubes used to fire the larger shells. The smaller shells are fired from racks

*A shell being lowered into a mortar. The large 10-inch shell in the fore-ground was fired last to end the display.*

of cardboard mortars mounted in wooden frames and held in position by wooden stakes driven into the ground. One of the last jobs was to place the shells in the mortars and to erect the set pieces.

By early evening the fireworks display was almost ready. Then it

*A workman inspects one of the set pieces before it is lifted into an upright position.*

happened—something that can stop a fireworks display dead in its tracks. A summer thunderstorm suddenly moved in and drenched the fairgrounds with more than an inch of rain. The display had to be postponed to the next evening.

Fortunately Bill and his crew were able to cover most of the fireworks with plastic sheets before the rain did much damage. Some of the set pieces had to be rebuilt, however, and the paper covering on a few of the shells got wet and had to be replaced.

The next night, promptly at 9:15 P.M., the display started with a volley of shells that filled the sky with dazzling color and earth-shaking explosions. Bill was in charge of setting off the shells, using a truck flare to do so. He had carefully worked out the firing order in advance and knew exactly when to light some shells or to radio Chris to ignite the fuse on a set piece. The set pieces at this year's celebration included two American flags, a sign spelling out WRAL

*An exploding shell fills the nighttime sky with streamers of color over the North Carolina State Fairgrounds.*

101, the call letters of a local radio station, and another honoring the Raleigh Parks and Recreation Department. The show lasted almost a half hour and ended with a giant shell, some ten inches across, which produced an array of color that could be seen for miles. As is the custom at fireworks displays, Bill had saved the best until last.

# 7

## Other Uses of Fireworks

THE BOY AND HIS FATHER sat quietly in the orchard, waiting for the sun to set. The boy glanced occasionally at the huge string of large firecrackers that hung from a nearby limb. Each of the fuses had been twisted around a long fuse, so that each firecracker hung separately to the side. Perhaps fifty crackers were tied to the center fuse.

The two people were not waiting for a Fourth of July celebration to begin. They were waiting for a flock of birds that usually came to the orchard each night to rest and feed on the ripening fruit.

As the sun dipped toward the western horizon, a dark cloud appeared in the east. The cloud swirled as it moved rapidly toward the waiting man and boy. The birds were coming.

As the huge flock circled the orchard, the man got up very slowly. He pulled a box of wooden matches from the bib of his overalls. With one of the matches, he carefully lit the end of the long fuse. Then he and his son quickly ducked behind a tree.

It seemed to the boy that it took a long, long time for the first firecracker to go off. Its noise was quickly followed by a second explo-

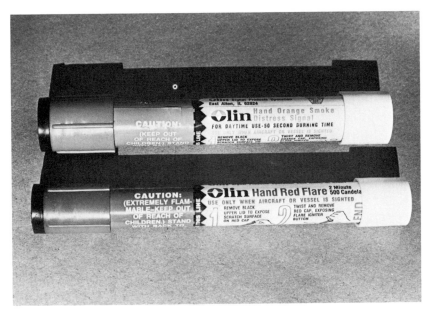

*The upper flare shown here gives off orange smoke for daytime use, while the lower one makes a bright red light for nighttime use.*

sion, and then a third. Then the whole orchard was filled with the sound of bursting firecrackers.

The birds, circling for landings in the trees below, were startled by the sudden burst of sound. The explosions went on and on as each of the fuses burned down to the powder. The noise echoed back and forth along the rows of trees. The birds wheeled in confusion, calling cries of danger to each other. Finally, the flock swooped upward and, after a moment, flew off toward the setting sun. The crop had been saved, for that night, at least.

Fireworks have been used for many generations in many countries of the world for this purpose. In England, explosives were combined with a mechanical scarecrow and called an "automatic crop protector." It was made so that a firecracker exploded every few seconds.

This explosion made three metal arms wave up and down. The explosions, along with the clatter and motion of the arms was "guaranteed to rid your field of birds and animal pests."

There are many other types of fireworks that are used in a variety of important ways other than in celebrations. Have you ever seen the red flares that trucks, buses, trains, and many cars carry? If the vehicle breaks down or has an accident, these brightly burning fountains are lighted and placed along the roadway to warn other traffic of the problem ahead.

Airplanes may carry flares attached to small parachutes. These fireworks burn with a very bright, white light. If the airplane is in trouble and needs to land at night, the flare can light up the ground for a long period of time as it slowly drifts toward the ground.

*A Very pistol is used to launch distress signals. The shells produce white or colored balls of fire.*

Ships have used rockets and Roman candles as signals for over a hundred years. Even in the days of radio, a ship in trouble may signal for help by shooting fireworks into the sky. These bright, colored lights help rescue ships find the sinking one in the dark.

In 1981, the U.S. Coast Guard began requiring boats operating in large bodies of water to carry emergency distress kits. These kits usually contain a hand-held rocket launcher called a *Very pistol*. The kit may also contain flares that give off bright lights or orange-colored smoke.

Fireworks have been used in warfare since their invention. *Star shells* are very bright flares that are attached to parachutes. They are similar to the flares used by airplanes, except that they can be fired from the ground from a special gun.

Since World War II, rockets have been used as major weapons of war. These and other missiles are simply nothing more than large fireworks carrying high explosives.

# 8

# Photographing and Enjoying
# Fireworks Displays

FOR MOST PEOPLE, there is nothing more exciting than watching a sky full of exploding fireworks. And many of us are sorry when the show is over and we wish we could carry something more than a memory home with us. With photography, you can do just that.

All that is needed to take beautiful photographs of fireworks is a camera that can take time exposures and some film for it. A tripod is also useful, although some interesting pictures can result from *not* holding the camera steady.

When using a single lens reflex camera, a mirror pops up when you take the picture. This mirror keeps you from seeing the burst of fireworks you are trying to photograph. If you hold the camera very steady while taking the picture, this should not be too much of a problem. However, a twin lens reflex camera or a camera with a separate viewfinder makes the job easier.

Use film that is as fast as possible because it takes less light to expose it. You can tell the speed of the film by looking on the box

*All you need to take pictures of fireworks is a camera and film for time exposures.*

for the ASA/ISO number. The larger the number, the faster the film and the less light required to get a good picture.

If you want color pictures or slides, try to get a film that is rated at least as fast as ASA/ISO 100. ASA/ISO 400 film is available in most stores and works best. If you want black-and-white pictures, use Tri-X film, which also has an ASA/ISO number of 400, or Plus-X which is slower with an ASA/ISO of 125.

The shutter should be set on Time or Bulb. The shutter will open when you push down on the shutter button and close when you release it, allowing you to make time exposures of varying lengths. The diaphragm, which controls the size of the beam of light entering the camera, should be halfway open, or set on f/8. The lens should be focused for photographing distant objects, or set on infinity.

In photographing fireworks, it is important to find a spot which

*A tripod helps make photographing fireworks an easier job.*

is safe and from which you can see the burst clearly without the danger of having someone step in front of your camera at the wrong time. A place as nearly directly under the explosions as possible is usually best. As soon as you have a solid place to rest your camera, you are ready to shoot. Lie on the ground, or sit so you can brace your camera by placing your elbows on your knees. Or rest your camera on some firm support, or put it on a tripod.

Because the night sky is black, the only image that will be recorded on your camera film is the light from the fireworks. But this light may not be as bright as you think it is. Therefore, you will need to keep your camera lens open for quite some time in order to get an image on the film. This is why a time exposure is needed.

Watch a few of the rockets as they go up until you can estimate about where they explode. Aim your camera at this area of the sky and watch for the trail left by a rising rocket. When the fireworks reach their highest point, they will seem to stop for a moment. Open your camera lens at this time and leave it open until the burst is fully open. Then close your camera lens.

This will probably give you a picture that shows the burst of light at its brightest. If you held the camera steady and closed the lens quickly enough, you will have a picture of hundreds of bright points of light.

*A picture like this one results when you don't use a long enough exposure.*

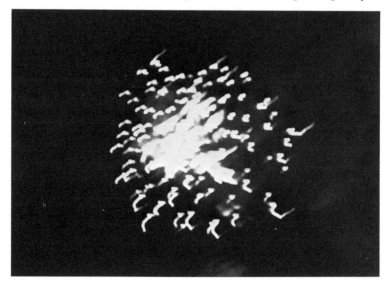

*With a little practice, you should be able to get pictures like this one.*

You can also take some interesting pictures by leaving the lens open until the burst is completely over. This will give you long streaks of colored light made as the burning parts of the fireworks fall back toward the earth.

Or you can intentionally move your camera while the lens is open. In this way, you can make swirls or streaks of light on the film.

Stationary displays can also be photographed. Again, fast film of ASA/ISO 100 or more must be used. Set your lens opening at f/8 and use an exposure of about 1/30 of a second. Since you have no way of knowing exactly how bright the light from the fireworks will be, these numbers may not always be accurate. It is suggested that you take several shots of the same display, using both longer and shorter exposures.

If you can get on top of a tall building to shoot some of your fireworks pictures, you may get some very interesting shots. Sometimes, by using a tripod to hold your camera very steady, you can get some of the detail of the ground in the pictures. Because of being farther away, you may have to leave your lens open a longer time.

Unlike pictures taken when you can measure the light and tell your subjects to "stand still and smile," you can never predict how well a fireworks photograph will turn out. So take a lot of shots and try a lot of different exposures. Sometimes a picture in which you thought you had made a mistake turns out to be the most interesting.

In addition to taking pictures, a fireworks display is a good place to observe a lot of science in action. Have you noticed how you see a burst of light a bit before hearing the explosion? This is because the light from the exploding fireworks travels to our eyes at the speed of 186,000 miles per second, whereas the sound travels much more slowly at about 1,100 feet per second. Since the light reaches you almost instantly and the sound takes a while, you can estimate the distance by timing the sound. This is done by counting "one thousand and one, one thousand and two," and so on as soon as you see the flash. The time it takes to say "one thousand and one" is about a second. One second between flash and sound indicates the explosion took place about 1,100 feet away, two seconds means 2,200 feet, and so on.

You can also use this method of estimating distance with nature's fireworks the next time you see a lightning flash. Five seconds between the flash and hearing the thunder tells you the lightning is about 5,500 feet, or a mile, away.

The exploding shells are the most spectacular part of any fireworks display. See if you can detect how many different kinds of shells are being used. Do some combine brilliant colors with strange sounds? Do some look like comets, with long tails in the sky? Do others explode into three or more bursts and each of these in turn explodes into several fountains of color? And how many different colors can you observe? As you can see, there is a lot to look for while enjoying a fireworks display.

We would like to end this book on a personal note. We have enjoyed telling you about the important part fireworks have played in the history of the world. In like manner, we hope you have enjoyed learning more about how fireworks are made and how they work. On the other hand, stories of people being killed or injured and fires being started by fireworks are another part of the story. These accounts should be reminders that fireworks can bring both joy and tragedy. Which it is depends mainly on how we choose to use them.

# Laws Governing the Sale
# and Use of Fireworks*

States which ban all Class C fireworks:

| | | |
|---|---|---|
| Arizona | Minnesota | Ohio |
| Connecticut | New Hampshire | Rhode Island |
| Delaware | New Jersey | Vermont |
| Georgia | New York | West Virginia |
| Massachusetts | North Carolina | |

States which allow only sparklers and/or snakes:

| | |
|---|---|
| Colorado (sparklers) | Iowa (sparklers and snakes) |
| Florida (sparklers) | Kentucky (sparklers and snakes) |
| Illinois (sparklers) | Oregon (sparklers and snakes) |
| Maine (sparklers) | Wisconsin (sparklers and snakes) |
| Maryland (sparklers) | |
| Pennsylvania (sparklers) | |
| Utah (sparklers) | |

States which allow Class C fireworks as approved by state and local enforcing authorities, or as specified in state laws (includes the District of Columbia):

| | | |
|---|---|---|
| California | Montana | Texas |
| District of Columbia | Nebraska | Virginia |
| Idaho | New Mexico | Washington |
| Indiana | North Dakota | Wyoming |
| Kansas | Oklahoma | |
| Michigan | South Carolina | |

States which allow Class C fireworks:

| | | |
|---|---|---|
| Alabama | Louisiana | South Dakota |
| Alaska | Mississippi | Tennessee |
| Arkansas | Missouri | |

States having no fireworks laws, except at the county level:

Hawaii
Nevada

---

* As of February 1, 1982, based on information supplied by the American Pyrotechnics Association.

# Index

Accidents, caused by fireworks
  celebration of Hearst election to
    Congress, 45–46
  celebration of Treaty of Aix-la-
    Chapelle, Paris, 45
  explosion at Pennsylvania Fireworks
    Display Company, 52, 54
  injuries in the United States, 48
Adams, Abigail, 7
Adams, President John, 7
Alchemists, 16

Bombs, 25, 26–27

Camera for photographing fireworks,
    67–71
Cherry bombs, 26–27
Class C fireworks, 24–25, 49
Consumer Product Safety
    Commission, 48

Dynamite used to stop fires, 35–36, 40

Early fireworks
  China, 15–17
  Europe, 18–19
  Greeks and Romans, 17–18

Film for photographing fireworks,
    67–68, 71
"Fire pill," 16

Firecrackers, 24, 26–27, 38, 41, 44, 46
Fires, caused by fireworks
  celebration of Treaty of Aix-la-
    Chapelle, 42–44, 45
  Oyens, Iowa, 38–39
  Portland, Maine, 41–43
  Remsen, Iowa, 37, 40
  Spencer, Iowa, 32–37
Fireworks, celebrations
  Brooklyn Bridge opening, 12–13
  Boston Pops Orchestra concerts, 50
  Erie Canal opening, 13–14
  Fourth of July at Washington
    Monument, 50
  Hudson-Fulton celebration, 13
  inauguration of President Grover
    Cleveland, 6
  International Fireworks
    Competition, Monte Carlo, 50–51
  King and Queen of Belgium, visit to
    United States, 8
  Knoxville World's Fair of 1982, 14
  marriage of Princess Elizabeth of
    England, 9
  New York City Fire Department
    celebration, 12
  New York Mets baseball games, 50
  Panama-Pacific Exposition, 13–14
  Philadelphia Centennial Exhibition,
    12
  Queen Anne Boleyn's arrival, 19
  Strassburg, Germany, 9
  Victoria Bridge opening, 8
  Winter Olympics, 1980, 50

Fireworks, ingredients, 53
  binder, 22, 52
  charcoal, 15–16
  fuel, 22
  oxidizer, 21–22
  saltpeter, 15–17
  special effects chemicals, 22–24
  sulfur, 16–17
Fireworks, kinds
  bombs, 26–27
  cherry bombs, 25, 33
  "fire pill," 16
  firecrackers, 24, 26–27, 38, 41, 44, 46
  flares, 61, 65–66
  flash crackers, 26–27
  fountains, 28, 29, 31
  gerbes, 29, 31
  "Greek fire," 17–18
  Japanese parlor fireworks, 17
  lances, 10
  "liquid fire," 17–18
  M-80's, 25
  novelty fireworks, 18
  pinwheels, 29, 32, 44
  rockets, 26–27, 32, 44, 66
  Roman candles, 26–27
  salutes, 25, 56, 57
  set pieces, 10–12, 31, 59–62
  shells, 28–31, 44–45, 51–54, 59–62,
    72–73
  Silver Salutes, 25
  smoke bombs, 25
  snakes, 20–21, 23
  sparklers, 20–22, 32–33, 37
  star shells, 66
  wheels, 29, 31
  worms, 20–21
Fireworks, manufacturing of, 22, 49–56
Flares, 61, 65–66
Flash crackers, 26–27
"Forge of Vulcan," 12

Fountains, 28, 29, 31
Fuses, 26, 28, 30, 46, 61

Gerbes, 29, 31
"Greek fire," 17–18
"Green Men," 19, 26
Grucci, Felix, 50–51
Grucci, Felix, Jr., 50, 52
Grucci, James, 50, 52
Grucci family, 48–51, 54–56
Gunpowder, 15–18, 26–28, 30, 44–45

Handel, George Frederick, 44
Hearst, William R., 45–46
Holidays celebrated with fireworks
  Bastille Day (French Independence
    Day), 8
  Christmas, 8–9
  Day of Saints, 9
  Fourth of July (Independence Day),
    7
  Holy Week, 9

Illegal fireworks, 25–26, 48
Injuries in United States, 48

Japanese parlor fireworks, 17

King and Queen of Belgium, visit to
  United States, 8

Lances, 10
Laser beams, 14
Laws governing the sale and use of
  fireworks, 75–76. See also
  Regulations and laws regarding
  the manufacture, sale, and use
  of fireworks
Lightning flash, timing distance away,
  72
"Liquid fire," 17–18

M-80's, 25
Marco Polo, 17
Mortars, 28–30, 45–46, 59–60
*Music for the Royal Fireworks*
    (Handel), 44

New York Pyrotechnic Products
    Company, 49–56
Novelty fireworks, 18

Observing fireworks displays, 72
Ogatsu family, 49

Pennsylvania Fireworks Display
    Company, 52–53
Photographing fireworks, 67–72
Pinwheels, 29, 32, 44
Punk, 32
Pyrotechnics, 8, 55

Regulations and laws regarding the
    manufacture, sale, and use of
    fireworks, 24–26, 37, 40–41,
    47–48, 59, 66, 75–76
Rockets, 26–27, 32, 44, 66
Roman candles, 26, 27
Ruggieri brothers, 44–45

Safety precautions, 54–56
St. George fighting a dragon, 9–10
Salutes, 25, 56, 57
Set pieces, 10–12, 31, 59–62
Shearon, Bill, Jr., 58–62
Shearon, Chris, 59
Shells, 28–31, 44–45, 51–54, 59–62,
    72–73
Silver Salutes, 25
Smoke bombs, 25
Snakes, 20–21, 23
Sparklers, 20–22, 32–33, 37
Star shells, 66

Thunder, used to time distance to
    lightning flashes, 72

Uses of fireworks
    celebrations, 7–14
    distress signals, 65–66
    frightening birds, 63–65
    nighttime illumination, 65–66
    warfare, 17–18, 66

Very pistol, 65–66
Vitale Fireworks Manufacturing
    Company, 58–59